J 629.43
HER

J 14263

DATE DUE

NO 20 '87			
DE 1 8 '87			
NOV 1 5 1990			

J
629.43
HER
Herda, D.J.
Research Satellites

J 14263

Enderlin Municipal Library
Enderlin, North Dakota 58027

RESEARCH SATELLITES

D.J. HERDA

RESEARCH SATELLITES

FRANKLIN WATTS | A FIRST BOOK
NEW YORK | LONDON | TORONTO | SYDNEY | 1987

Cover photograph courtesy of NASA

Photographs courtesy of: Charles Phelps Cushing Collection/H. Armstrong Roberts: p. 12; NASA: pp. 15, 16, 26, 27, 29, 32, 34, 35, 40, 41, 43, 46, 50; Sovfoto: p. 21; National Oceanographic and Atmospheric Administration (NOAA): pp. 56, 59.

Library of Congress Cataloging-in-Publication Data

Herda, D. J., 1948–
Research satellites.

(A First book)
Includes index.
Summary: Traces the history and development of research satellites and discusses their uses in cartography, meteorology, and climatology.
1. Scientific satellites—Juvenile literature.
[1. Scientific satellites] I. Title.
TL798.S3H47 1987   629.43′5   86-24225
ISBN 0-531-10311-0

Copyright © 1987 by D.J. Herda
All rights reserved
Printed in the United States of America
6 5 4 3 2 1

CONTENTS

Chapter One
The Race for Space 9

Chapter Two
Destination: Moon 19

Chapter Three
Weather Watch 25

Chapter Four
Earth Watch 37

Chapter Five
Food Watch 45

Chapter Six
Ocean Watch 53

Chapter Seven
Growing Larger Every Day 63

Glossary 67

Index 69

RESEARCH SATELLITES

CHAPTER ONE
THE RACE FOR SPACE

"Four, three, two, one, zero. Ignition. *Lift-off*. We're clear of tower at zero minus three seconds. It's a go!"

"Space travel." The simple two-word phrase stirs images that extend into infinity—images of humans hurtling through space, of giant megarockets roaring between *planets*, of strange lifeless worlds . . . and worlds perhaps not so lifeless.

But when did space travel begin? Was it with the firing of the first rocket? Or the launching of the first *earth-orbiting* satellite? Or, perhaps, did it begin long, long before those events?

Before the turn of the century, a Russian schoolmaster by the name of Konstantin Eduardovich Tsiolkovski (Chawl-KOF-skee) had a dream. The year was 1895, before Orville and Wilbur Wright's historic first manned airplane flight at Kitty Hawk, before Henry Ford's first automobile, and before Thomas Alva Edison perfected the light bulb. In that year Tsiolkovski published his book *Dream of the Earth and the Sky*, in which the Russian educator first mentioned the possibilities of an artificial satellite.

"The fancied satellite of the Earth would be something like a moon," he wrote, "but arbitrarily closer to our planet—only far enough away to be outside its atmosphere, that is, at a distance of some 300 versts." (A verst is a Russian measurement of length of about 3,500 feet [1,067 m].)

Tsiolkovski never launched the satellite about which he dreamed. But until his death in 1935, he wrote extensively about the reality of rockets lifting "artificial moons" into the emptiness of space. His work motivated such pioneers as American Robert H. Goddard and Germans Johannes Winkler, Max Valier, and Wernher von Braun to experiment with space flight.

In 1930, the eighteen-year-old von Braun and a group of German physicists began working to develop rockets capable of lifting small payloads into space. Within a short time, Adolf Hitler, who had recently come to power in Germany, became interested in the experiments.

Although Hitler at first failed to grasp the full implications of rockets as a wartime tool, he soon came to regard them as critical to waging war and threw his weight and influence behind the German production of liquid-fuel rockets. During the last two years of World War II, the Nazi leader became obsessed with developing rockets with mighty payloads of explosives capable of flying faster than aircraft—rockets that could help him realize his goal of world domination.

Luckily, Hitler's goal of building a mighty arsenal of wartime rockets never materialized. The Allied Powers' defeat of Germany (and the other Axis countries, Japan and Italy) put an end to German domination of the skies. Following the war, many of Germany's rockets were shipped to the United States for study and experimentation, and many of Germany's leading scientists—including von Braun—came with them.

A new era of space exploration was opening up—one not geared to conquest and war but to discovery and peace.

But optimism about launching exploratory satellites into space was marred by reality. In 1946, the German A-4 (V-2) rocket, the leading rocket in design and performance, had demonstrated a top speed of 3,355 miles (5,400 km) per hour. To place an object into even a relatively low orbit, a rocket would need to travel in excess of 17,000 miles (27,359 km) per hour. Where would such a rocket come from?

That year, the U.S. Navy and the newly created Bureau for Aeronautics designed on paper a rocket capable of reaching such high speeds and injecting itself into orbit. But the estimated cost of constructing such a rocket was $8 million in 1946 prices—far more than the Navy could justify for an experimental project.

At about the same time, the Army Air Force—along with the privately funded RAND (Research and Development) group—undertook an independent study and, on May 12, 1946, unveiled plans for the "Preliminary Design of an Experimental World-Circling Spaceship"—a *multistage rocket* whose purpose would be "to collect information on cosmic rays, gravitation, terrestrial magnetism, astronomy, meteorology, and properties of the upper atmosphere." The proposed rocket had four stages and a maximum payload of 500 pounds (226.8 kg).

For various reasons—including the high cost of building such a rocket—plans for its construction were abandoned. Then, in 1951, the British Interplanetary Society unveiled plans for a rocket weighing less than 17 metric tons and capable of carrying a cosmic ray instrument and a radio transmitter into space, all at a cost far lower than previous estimates. Still, the United States had other places to spend its money—on strategic bombers and military missiles, for example. So the U.S. Department of Defense dismissed the plan as impractical.

But someone was listening.

On November 27, 1953, at the World Peace Conference held in Vienna, Alexander N. Nesmeyanov (Nez-MY-in-off) of the Russian Academy of Sciences announced to the world that "the creation of an artificial satellite of earth is a real possibility." Less than four years later, the United States and the world were shocked to learn that the Russians had been right. On October 4, 1957, the USSR announced the successful launch and orbiting of *Sputnik 1*, the first artificial earth satellite.

Sputnik 1 was an unlikely looking celebrity. It consisted of an aluminum ball some 22.8 inches (58 cm) in diameter and weighing 184.3 pounds (83.6 kg). Inside the ball were instruments for measuring *density* and temperature throughout the satellite's orbit, which ranged from 141 miles (227 km) to 585 miles (941 km) from

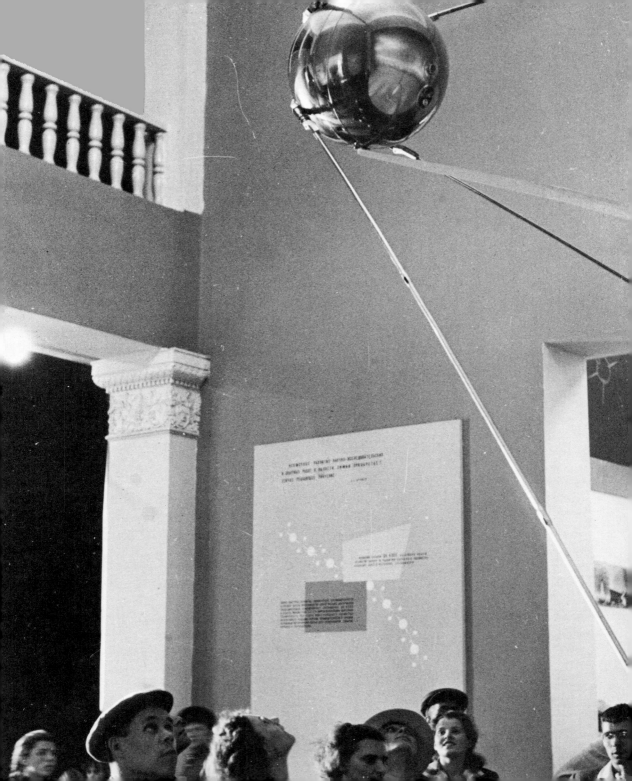

earth. It also contained instruments for collecting data on the concentration of electrons in the ionosphere. *Sputnik 1* was launched into space atop a modified R.7 rocket, the first Soviet intercontinental *ballistic* missile (*ICBM*).

Immediately following word of the successful Soviet launch of *Sputnik 1*, Wernher von Braun approached the Defense Department for permission to launch America's first satellite, which he insisted could be done within ninety days. But Defense Secretary Neil McElroy was reluctant to jump into the space race without adequate preparation and placed von Braun's request on hold.

Then, on November 3, 1957, the Soviets launched *Sputnik 2* into orbit with its 1,120-pound (508-kg) payload, including the first live passenger, a dog named Laika, and various instruments to monitor her condition in the *weightlessness* of space. *Sputnik 2* also contained instruments for measuring the radiation environment of space.

Again von Braun approached the Defense Department. This time, he and his team of aeronautical engineers at the Army Ballistic Missile Agency in Huntsville, Alabama, received approval for launching America's first satellite. Within three months, *Explorer 1*, America's answer to the Sputniks, was in orbit.

Explorer 1 hardly matched *Sputnik 2* in size or weight, measuring less than 40 inches (1 m) and weighing only 10.5 pounds (4.8 kg). But it carried both internal and external thermometers, erosion gauges, and an impact microphone for detecting the flux of micrometeoroids, plus a Geiger-Mueller counter for recording cosmic rays striking it. Data recorded by these instruments were stored on a miniature tape recorder, then "dumped" on command from earth as the satellite passed over one of several *tracking stations*.

Sputnik 1, *the first artificial satellite, was launched in 1957. Today it is housed in the Academy of Sciences in Moscow.*

The biggest breakthrough for *Explorer 1* was in the miniaturization of its instruments. They were small and lightweight enough to fit into a tiny package—something that would pay dividends to the United States in future launchings.

On February 5, 1958, the United States attempted to follow up its successful Explorer launch with the launching of the recently developed *Vanguard 1* satellite, but, like a previous effort, the attempt was a failure. A third attempted Vanguard launch on March 17 was a success, resulting in the second U.S. satellite in Earth orbit, a 6.4-inch (16.3-cm) ball weighing a scant 3.25 pounds (1.47 kg). The Soviet press, unimpressed by the accomplishment, ridiculed the satellite by referring to it as a "grapefruit" because of its small size.

Over the course of the next two years, eight more Vanguard launchings were attempted; only two succeeded. However, even in failure there is sometimes success. Both the satellites and their Vanguard rocket launch vehicles contributed to space science and aerospace technology.

The Vanguard satellites proved that *solar cells* could supply electrical power for the functioning of on-board instruments. They also gave geophysicists—scientists who study the physical size and properties of the earth—a truer picture of the earth's shape and makeup, resulting in a more precise mapping of islands in the Pacific Ocean and in laying the foundation for future meteorological satellites.

The Vanguard launch vehicles led to the design and engineering of larger, multistage rockets necessary to lift additional satellites with ever-larger payloads into orbit.

By 1961, the United States was regularly relying on such launch vehicles as *Juno I* and *Juno II* and the more powerful Atlas and Thor

One of the Vanguard satellites—notable for their miniaturized instruments

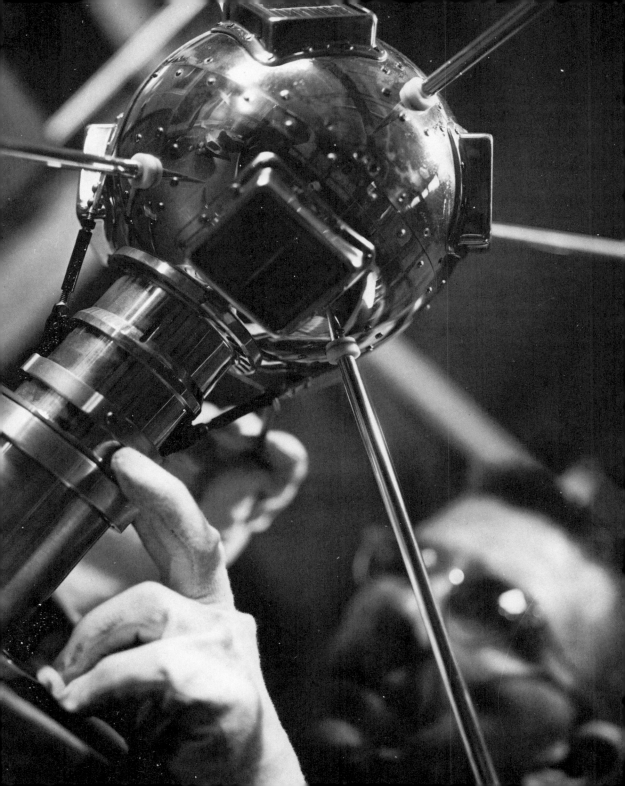

rockets for lifting hundreds and even thousands of kilograms of weight into orbit—a far cry from the minuscule *Vanguard 1* satellite.

Meanwhile, as the technology behind its launch rockets grew, the USSR was lifting ever-heavier scientific satellites into orbit. Soviet launch rockets were substantially larger and more powerful than the largest U.S. vehicles. With its stable of reliable, heavy-duty rockets, the USSR quickly took the lead in the early race for space.

But the race had just begun.

Vanguard 2 *rests atop its launching vehicle, its protective nose cone yet to be put in place.*

CHAPTER 2
DESTINATION: MOON

In the late 1950s, the idea of sending an unmanned satellite to the moon seemed a reasonable goal. Now that Sputnik and Explorer had proven our ability to launch satellites beyond earth's atmosphere, sending a payload to the moon seemed a relatively small challenge to meet. A little more launch power, some additional navigational technology, and the task should be done.

On August 17, 1958, amid much excitement, the United States launched its first scientific *probe* (an unmanned vehicle sent into space to gather information) from Cape Canaveral, Florida, toward the moon. The launch vehicle was a powerful Thor-Able rocket, which was expected to perform flawlessly.

And so it did—at first. But just seventy-seven seconds into the flight, an explosion in the first stage of the vehicle brought a sudden halt to the mission. The probe, nicknamed *Pioneer 0,* had traveled only 1/27,000 the way to the moon—a distance of about 10 miles (16 km). Clearly the United States would have to do better on its next attempt.

Undaunted by *Pioneer 0*'s failure, the United States launched *Pioneer 1* on October 11, but it, too, failed. On November 8, still another moonbound missile went astray, although *Pioneer 2* managed to achieve an altitude of 963 miles (1,550 km) before faltering, considerably better than its two predecessors.

On December 6, 1958, *Pioneer 3* was launched aboard a *Juno II* vehicle. Although an early shutdown of the first stage of the rocket doomed the probe to failure, the vehicle reached an altitude of 63,580 miles (102,300 km), transmitting valuable data to earth as the probe passed through the *Van Allen radiation belt* and again as it plunged back to earth.

The Soviets, meanwhile, were eyeing America's *interplanetary probes* with interest. They, too, wished to be the first nation to place a data-collecting probe in the vicinity of the moon, and the recent flurry of American activity had them concerned.

On January 2, 1959, the USSR launched the first artificial object to escape the gravitational pull of the earth and take up a *trajectory* that would carry it close to the moon and eventually onward into orbit around the sun.

Luna 1 was launched from the Baikonur space center by a modified version of the same vehicle that sent *Sputnik 1* into orbit. For the *lunar* mission, the Soviets used an SS-6 *Sapwood* ICBM fitted with an additional stage that would permit a payload weighing as much as 882 pounds (400 kg) to escape earth's gravity.

Luna 1 was probably meant to collide with the moon, sending back data until the moment of collision, but it failed to do so, passing instead some 3,700 miles (5,955 km) from the surface of the moon on its way to a solar orbit.

The *Luna 1* probe was fitted to measure solar and cosmic radiation, interplanetary magnetic fields, and gaseous composition of the regions through which it traveled. It was also equipped with a system for releasing vaporized sodium. When the probe reached a point 70,000 miles (112,630 km) from earth, a cloud of gas was ejected from the rocket stage attached to the probe. Solar radiation caused the cloud to glow, and tracking cameras on earth photo-

The first artificial object to escape Earth's gravitational pull, Luna 1 *passed close to the moon's surface, transmitting data back to Earth.*

graphed it against the constellation Aquarius. The gas provided scientists on earth an opportunity to study the probe's trajectory and calculate the density of matter in interplanetary space.

Meanwhile, the United States was preparing the last of its Pioneer probes for launching, and on March 3, 1959, it was sent on a similar mission to that of *Luna 1*. The 13.2-pound (6-kg) *Pioneer 4*, launched atop a *Juno II* rocket, was programmed to fly past the moon at a distance of about 15,000 miles (24,135 km). However, the second-stage *solid-propellant* rockets burned for a second longer than planned, throwing the lunar probe some 22,000 miles (35,405 km) off course as it flew by the moon at 37,300 miles (60,016 km) on its way into solar orbit.

Although tracking stations lost contact with the probe when it reached 407,000 miles (654,863 km) from earth, *Pioneer 4*, like *Pioneer 3* before it, provided valuable data on the Van Allen radiation belt by means of a Geiger-Mueller counter specially designed for the purpose.

Although *Luna 1* and *Pioneer 4* added to our knowledge of interplanetary space, the greatest success of the decade went to Russian probes *Luna 2* and *3*. *Luna 2* was launched on September 12, 1959. It weighed a whopping 860 pounds (390 kg) and was filled with instruments to detect and measure magnetic fields and radiation from the moon. The final stage of the launch vehicle, which accompanied the probe to the moon, also released a sodium cloud that was observed from earth.

Luna 2 made a planned crash landing into the moon at a point east of an area known as the Mare Serenitatis, leaving at the crash site a specially designed insignia of the USSR and the landing date.

Luna 3 was launched on October 4, 1959, by the same type of vehicle that had launched the two previous Lunas. This probe weighed 614 pounds (278.5 kg) and was a scientific and engineering success, demonstrating to the world that the Soviets were quickly perfecting the technology for producing complex interplanetary spacecraft. This third Luna probe was designed to circle the moon and photograph that part of the sphere always hidden from earth.

The satellite performed flawlessly, photographing the dark side of the moon for some forty minutes from an altitude of 4,900 miles (7,884 km) before swinging back toward earth to telemeter its pictures back through space to a receiving station in the Soviet Union.

The photographs from *Luna 3*, pieced together like a jigsaw puzzle by Soviet scientists on earth, revealed fewer and smaller *maria*—dark planes on the surface—than on the visible side of the moon, plus many large craters and clusters of medium-size ones. Altogether, *Luna 3* photographed a remarkable 70 percent of the side of the moon never before seen.

With the data received from *Pioneer 4* and the Luna probes, plus the crash landing of the first probe on the moon, humans had taken their first steps toward opening the frontiers of space. Attempts at collecting data from these early research probes were crude by present standards, but they were enough to fuel the imagination and whet the appetite for an ever-increasing array of information about the vastness of space.

The big question remaining to be answered as the 1950s came to a close and a new decade began was simple: Where would we go from here?

CHAPTER THREE
WEATHER WATCH

Once the United States and the Soviet Union had successfully placed scientific satellites in orbit, the next step was to develop practical applications for future satellites. What kind of data could they furnish and how could they be used on a regular basis to improve the quality of life on earth?

Early successes from orbiting probes and data-gathering equipment soon caught the eye of meteorologists (scientists who study the earth's atmosphere as it affects weather). It seemed very possible that they might obtain more accurate information regarding the world's ever-changing weather systems from data relayed to earth from orbiting satellites. What better vantage point for studying a developing storm than from hundreds of miles above the earth's surface?

As early as October, 1954, a *Vanguard 2* satellite, launched from White Sands Proving Grounds in New Mexico, photographed an otherwise-undetected storm. Six years later, the United States took the initiative in creating an interplanetary weather-watch system by developing a family of satellites called TIROS (television and infrared observation satellite). The first TIROS was launched by *NASA* (National Aeronautics and Space Administration) into orbit on April 1, 1960. During the next five years, nine additional TIROS satellites were successfully launched, each carrying a pair of miniature television cameras and many equipped with *infrared radiometers* and other radiation-sensing equipment.

Left: this TIROS meteorological satellite is 22 inches (56 cm) high and weighs almost 300 pounds (136 kg). Underneath you can see two independent television camera systems. Each camera consists of a vidicon tube and a focal plane shutter which permit pictures to be stored on the tube face. An electronic beam converts a "stored picture" into a TV-type signal which is radioed to a ground station for read out.
Above: a simplified schematic of a TIROS satellite.

The early successes of the TIROS satellites led to the development and deployment of other families of weather-detecting satellites. These new satellites brought a series of cameras and sensing devices into routine operation for both day and night observation of cloud patterns. The satellites also prompted the development of equipment capable of determining temperature, cloud heights, wind velocity and direction, and other environmental conditions.

TIROS-N ■ These satellites, an outgrowth of the original TIROS satellites, benefited from experience in the design of NASA's original TIROS series of ten research satellites.

TIROS-N satellites carried into orbit the Advanced Very High Resolution Radiometer (AVHRR), designed to increase the amount of radiometric information for more accurate sea-surface temperature mapping and the identification of areas of snow and sea ice. A unique Data-Collection System (DCS) received environmental data from fixed and moving platforms such as buoys and balloons and stored it for transmission to ground stations at a later date. A solar environmental monitor was included to measure proton, electron, and alpha particle densities for predicting solar disturbances that affect the earth's weather.

ESSA ■ In 1966, the first ESSA (Environmental Science Service Administration) operational satellite was launched, followed by eight additional satellites. (ESSA is now the National Oceanic and Atmospheric Administration [NOAA].) Later ESSA models carried two advanced vidicon cameras (AVCs), each capable of complete daily picture coverage of the earth's weather. ESSA satellites used large-format television cameras of 1 inch (2.54 cm) in order to provide improved image quality of global cloud cover over that relayed to earth by the smaller-format cameras on TIROS satellites, which used a ½-inch (1.27-cm) format.

ITOS ■ Improved TIROS operational satellites (ITOS) were launched by NASA in 1970–71. ITOS satellites were capable of providing global observation of the earth's cloud cover every twelve hours as compared to every twenty-four by ESSA satellites. In addi-

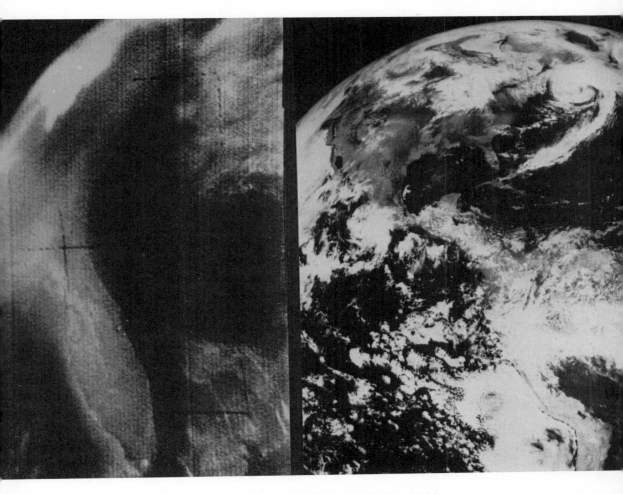

From the launching of the first TIROS satellite in 1960, refinements in the satellite's data-gathering ability were continually being made. Notice the contrast in quality of an image transmitted from a TIROS satellite in 1960 and one in 1975 from a greater distance.

Enderlin Municipal Library
Enderlin, North Dakota 58027

tion, ITOS carried new instruments capable of providing day-and-night imaging by means of very-high-resolution radiometers (VHRRs). They were also equipped with vertical temperature profile radiometers (VTPRs) for taking temperature readings of the atmosphere.

Advanced TIROS-N ■ Most recently, Advanced TIROS-N satellites, some measuring up to 14 feet (4.3 m) in length and weighing up to 3,800 pounds (1,724 kg), have joined other families of satellites circling the globe. Advanced TIROS-N satellites are equipped with search-and-rescue (SAR) antennas, which provide data for locating and identifying ships in trouble or aircraft which may have come down in remote places. Such information can then be used to help rescue squads reach troubled vessels.

Of course, while the United States was launching a wide range of weather satellites, the Soviet Union was following suit. A series of Soviet Meteor satellites was successfully launched in 1969 and continue to provide weather data for Soviet consumption today. Meteor satellites routinely scan the earth from one pole to the other, observing an area of 11,584 square miles (30,000 sq km) per hour.

The Soviet system provides routine forecasts of weather phenomena like fast-developing tropical storms. Armed with such forecasts, the USSR can act to ensure its interests against possible disaster.

The Soviet weather satellite system is also valuable in predicting snowmelt from various Soviet mountain ranges and in developing irrigation plans for dry, remote places, which are numerous in certain regions of the Soviet Union. It also aids Soviet commerce by helping to reroute shipping to avoid major storms, rough seas, high winds, and ice.

Once regarded as curiosities, dozens of orbiting research weather satellites are being used regularly today by meteorologists and environmental scientists on a worldwide basis. Since 1966, the entire earth has been photographed by satellite-borne cameras on a continuous basis at least daily. The photos are indispensable for analyzing developing weather patterns, as well as for making both short- and long-range predictions.

In the United States, meteorological information from satellites is received for processing and worldwide distribution at the National Environmental Satellite Service (now called the National Environmental Satellite, Data, and Information Service) in Washington, D. C. Some of the information is nearly impossible to obtain in any other way—readings taken from the oceanic areas of the Northern and Southern hemispheres, for example, as well as from deserts and the North and South poles.

Orbiting satellites locate large-scale cloud formations that may reveal storm systems, fronts, upper-level troughs and ridges, jet streams, fog, stratus, sea-ice conditions, snow cover, and upper-level wind directions and speeds.

Data received from satellites are also useful in tracking the courses of hurricanes, typhoons, and tropical storms. Coastal and island stations with little knowledge of what's happening at sea a scant hundred miles away rely on satellite data for information about the presence and position of frontal patterns, storms, and general cloud cover.

Of course, weather satellites don't tell meteorologists everything they need to know about the earth's weather. When a major hurricane is found to be heading in the direction of the United States, for instance, *USAF* (United States Air Force) and *NOAA* (National Oceanic and Atmospheric Administration) weather-reconnaissance aircraft are summoned to support the satellite data and to provide more detailed measurements of meteorological events in and around the storm.

When a hurricane moves close to the United States, coastal radar stations keep the storm center under close scrutiny. In this way, precise information on the path and strength of a storm is passed along to communities that may lie along its path.

Other sources of data include oceangoing ships, buoys, and weather stations located on various islands near the path of major storms.

In this picture transmitted from a weather satellite over the Caribbean, a swirling hurricane moves westward towards Central America. Cuba and, above it, Florida are clearly visible.

Since the inauguration of daily satellite coverage of the earth in 1966, no tropical storm has escaped detection and daily tracking. Most storms are detected while still developing, often at distances beyond the normal range of reconnaissance aircraft.

To most of us, the value of weather satellites is obvious from the charts and maps that appear nightly on our favorite television stations. But the United States and the Soviet Union are not alone in developing weather satellites. A lot of important work is being done by various European countries. In addition, Japan, India, and China have recently deployed weather satellites in order to provide advance warnings of hurricanes, tornadoes, monsoons, and flood-producing rainstorms, helping to save hundreds of thousands of lives and billions of dollars each year.

Yet, as important as weather satellites are to the world today, they are only a small part of the network of earth-watch satellites providing detailed data on the planet on which we live . . . on which we all depend.

Satellites come in all shapes and sizes: (a) the tiny Explorer satellites, each equipped with an octagonal platform and four solar paddles; (b) IRAS, the Infrared Astronomy Satellite, had delicate light sensors; (c) Mariner 2, with two television cameras and equipment to study infrared and ultraviolet radiations, magnetic fields, and solar particles; (d) Voyager 2, weighing almost a thousand pounds and equipped with TV cameras and ultraviolet and infrared sensors.

(a)

(b)

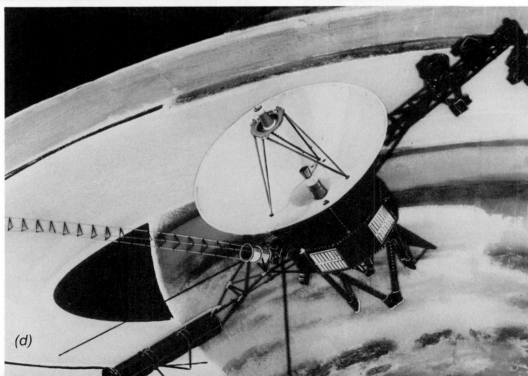

(d)

CHAPTER FOUR
EARTH WATCH

Soon after the opening of the space age, we became aware of the value of satellites for monitoring the earth's natural resources. Our first hint came in 1960 when TIROS weather satellites showed maplike outlines of the world beneath the clouds.

These early satellite images went largely unnoticed until someone spotted a pattern in the snow where northern Canadian lumberjacks had cleared the forests.

Still, it wasn't until humans went into space that we came to realize the immense power of satellite-generated images. Astronaut Gordon Cooper, during his flight in a Mercury space capsule in May 1963, astonished space officials by claiming to see roads, buildings, and even smoke from chimneys here on earth. Mission Control accused him of hallucinating.

But Cooper wasn't hallucinating. And closer examination of the film returned from Mercury and Gemini space flights showed not only what Cooper had seen, but also such changing landmarks as urban structures, the building of new roads, drainage patterns in western Texas, and wheat fields in Kansas.

Other space photos showed where rain had fallen the previous evening in a dry region of Texas, not because the ground was visibly damp, but because the vegetation had begun to "uncurl," throwing up a different-color response to the camera.

Several *Apollo 9* photos showed snow concentrations on a mountain range in Arizona, as well as extensive flood damage in an area of some 165 square miles (427 sq km) along the Ouachita River in Louisiana.

Before long, scientists had developed new techniques to improve the value of satellite observations of the earth. Information was obtained through multispectral imagery in visible light and infrared (IR), which made possible the detection of tiny variations in radiation on the ground. Two different types of instruments were used—infrared cameras loaded with IR film, and infrared radiometers, which pick up only IR wavelengths.

Early IR photographs taken by research satellites showed the difference between fields containing healthy agricultural crops and those afflicted with blight. The healthy crops registered red while the blighted ones registered black. Often, satellite photos identified the onset of crop diseases even before they became obvious to the farmer on the ground.

Today, multispectral sensors are commonly used to identify different features on the earth's surface. They record the different types of energy—called spectral energy—emitted or reflected from the sun. The spectral energy emitted by a plant is quite different from that emitted by a rock or water, for example.

These differences are recorded by sensing equipment aboard the satellite and then interpreted in different colors to display the variations within the region surveyed. Rocks may appear red, for instance, while trees and shrubs appear orange or purple. This type of reconnaissance has become so effective that today it's possible to distinguish between specific crops such as beans, peas, and corn. Changes in the spectral signature—the amount of radiation given off—also indicate such information as soil condition, moisture content of the soil, and crops affected by disease or insect infestation.

The first satellites designed to concentrate on defining the condition of the earth were developed by General Electric for NASA. Called Landsats, they operate in north–south circular orbits at an altitude of about 570 miles (917 km). Their coverage of land and ocean moves progressively from east to west as the earth rotates beneath them. Their orbits enable a single satellite to report on nearly every area in the world every eighteen days.

Landsat images are digitized—changed to electronic impulses like those of a computer—and transmitted to dish antennas at receiving stations on earth. There the images are stored on magnetic tape until they're turned into photographic prints in color and black and white.

Landsat 4, the fourth in the Landsat series of satellites, was designed to survey the earth from an altitude of more than 398 miles (640 km), using improved sensing equipment. In addition to carrying a multispectral scanner (MSS) of the kind used on the first three Landsats, *Landsat 4* carried a *sensor* known as a thematic mapper (TM) capable of differentiating between features as small as 0.2 acre, as compared to 1.2 acres previously. This enabled scientists and technicians to extract much more detailed information from a typical photograph.

Landsat 4 was unique because it was the first earth-watch satellite designed to be retrieved by the space shuttle for possible repair and reuse. It combined a general-purpose satellite "bus," supplying the basic power, propulsion, *attitude control*, communications, and data handling required of a satellite, with a structure capable of accepting a broad range of scientific instruments.

From all the data being transmitted to receiving stations on earth, we are finally beginning to learn what the earth on which people have lived for millions of years actually looks like. We have discovered that it isn't round or even egg-shaped, as we had previously thought, but is actually more pear-shaped. This and other new data have helped us to create more accurate maps of the earth's geographical features.

Before the era of satellite observation, many areas—even in well-developed parts of the world—had been incorrectly mapped. Landsat imagery helped correct and update certain features of existing U.S. maps at scales of 1:250,000 and larger.

Satellite imagery has been very beneficial in producing detailed maps for the construction of roads, railways, and irrigation chan-

ABOVE: Landsat 4, *the fourth in the Landsat series of satellites, carried a sensor known as a thematic mapper, capable of differentiating between features as small as 0.2 acres.*

LEFT: a Landsat undergoing a final check before launching. Visible at the top of the spacecraft is the Tracking and Data Relay Satellite System (TDRSS) antenna. When in space orbit, this antenna is extended 12.5 feet (4 m) above the body of the satellite.

nels, saving untold time and countless millions of dollars in construction costs. It has also made possible the charting of underwater features such as coral reefs, resulting in less danger to shipping.

In the USSR, imagery obtained from Salyut space stations similar to Landsat satellites proved invaluable for checking the route of the BAM strategic railway, then under construction north of the existing trans-Siberian rail route.

Before Landsat, more than half of Asia, Africa, and Latin America had not been mapped adequately at scales larger than 1:1,000,000. Since Landsat, uncharted areas have been mapped quickly and inexpensively at scales of 1:250,000, about four times larger than possible before, and existing maps have been updated with greater accuracy. Landsat has also helped pinpoint those geographical features requiring greater accuracy and higher resolution than possible to achieve through aerial reconnaissance by conventional aircraft.

How does Landsat's ability to provide detailed, accurate maps of an area affect us on a day-to-day basis? Imagine that you are the head of the engineering department for a major oil firm. Your job is to locate and drill a field of oil deposits in the jungles of South America. From the maps you have on hand, you determine a likely location for mass deposits of valuable oil some 1,800 miles (2,900 km) into the jungle.

But when you arrive with your crew and equipment, you discover the area of flatland you expected to find is actually the peak of an 1,800-foot (2-km) mountain. Drilling through it is impossible. You've lost countless months of work, hundreds of hours of labor, and several hundred thousand dollars in expenses.

In desperation, you obtain a copy of a Landsat map and discover a more likely drilling site only 6 miles (10 km) to the east. Within three months, you're pumping oil for your company.

Or perhaps you're a geographical cartographer hired by your government to produce a new map of Egypt at a scale of 1:1,000,000. You begin the 10-year project with an estimated bud-

Memphis, Tennessee, on the Mississippi River, as seen from Landsat 4. *Landsat imagery has helped correct and update existing U.S. maps.*

get of $2.4 million, using black-and-white aerial photos taken over the last eighteen months. But there are many geographical irregularities you can't determine because of the lack of data in many of the photographs.

Finally, you turn to a Landsat map and find three times the detail as you had received from the aerial photos. The completeness of the maps and the speed with which space surveys can be completed allow you to complete your task in half the time for far less money.

Both cases are based on actual incidents. They are just two examples of how Landsat satellites make life simpler, less costly, and more efficient in providing detailed, accurate maps of the surface of the world on which we live.

But cartography, or mapmaking, isn't the only area in which earth-watch research satellites benefit us. Just as they help us to detail its surface features, so, too, do they help us to provide food for ourselves.

CHAPTER FIVE
FOOD WATCH

In the mid-1970s, NASA, the Department of Agriculture, and NOAA combined forces in a major experiment to demonstrate how satellite monitoring could help forecast the yield of a crucial world food crop—wheat.

The researchers calculated the total wheat acreage from Landsat surveys and compared the results with the potential yield per acre based upon past meteorological data. The survey turned out to be remarkably accurate and was later expanded to include additional crops.

In another survey, researchers relying on Landsat images of California's Imperial Valley succeeded in identifying more than twenty-five separate crops in 8,865 fields. The total area covered was 643 square miles (1,664 sq km). Among the crops picked out were corn, popcorn, sorghum, oats, soya beans, grasses (rye, Alicia, Sudan, and Bermuda), lettuce, mustard, carrots, tomatoes, and onions. The survey took little more than forty hours.

Researchers were also able to distinguish between wet-planted fields and bare earth in areas as small as ten acres (4.5 ha). Newer Landsat satellites are capable of identifying crops on plots as small as one acre in size.

The ability to distinguish between crops in Landsat images may eventually result in a form of global food watch that could help us avoid disastrous food shortages. Such a program could yield benefits to the United States of $200 million annually and eliminate

A Landsat 4 *thematic-mapper image of South Central Iowa. The rectangles are corn fields. The original color photograph showed fields of growing corn as red, fallow fields as blue-gray, and wet fields as dark blue to black. A cirrus cloud and its shadow can be seen in the upper part of the image.*

many of the problems brought about by the under- and overproduction of various crops.

Recent additional research points to the possibility of achieving better management of crop resources. From a regular survey made by satellites, we can discover the best time to plant and harvest various crops for highest yield based on soil condition and moisture content. An inventory of crops can be kept during the growing season and advance warning can be given of drought, floods, and erosion.

A similar type of food-watch program, in combination with high-speed mainline computers on the ground, would allow an entire region of farmland to be sorted crop by crop in a matter of hours. The result would be a computer-printed map showing the exact location and area of each crop with an accuracy of better than 90 percent.

One example of how satellites are currently working to overcome global food shortages involves the recent severe drought in the Sahel, a vast African prairie just south of the Sahara.

Several countries in the area asked to have the Sahel surveyed by Landsat in order to discover ways of alleviating starvation due to drought. Landsat not only identified many regions as being suitable for farming, but also pointed out a valuable lesson in conservation.

The satellite found an irregularly shaped area of vegetation that stood out from surrounding parcels of parched land. The patch was identified as a ranch where careful management had prevented the livestock from denuding the land and turning it into worthless desert.

By properly rotating grazing livestock and using other methods of livestock management, humankind can prevent the encroachment of the desert and the devastating dust bowls that often follow.

Landsat helped the people of the Sahel to determine the best times to turn out their cattle in grazing areas. Plans were prepared to prevent overgrazing and to open up new areas to cattle. The satellite also helped determine how much additional investment was needed in land improvements for draining, irrigation, seeding, and the application of fertilizers.

Along similar lines, Landsat data have shown clear advantages in estimating the volume of timber across large areas of a country. Satellite images can offer data regarding the best time to cut timber, the proper amount to cut, and the best conservation techniques to reclaim cutover land.

Developing countries are finally becoming aware of the need to manage their forest resources, not only as a means of meeting their present timber requirements, but also as a way of preserving the ecological balance and preventing soil erosion, the silting of dams, and the polluting of coastal waters—all of which adversely affect the ability of an area to produce food for its people.

In Brazil, Landsat has been used to monitor a program for controlled development of the Amazon's forests for commercial purposes, including cattle grazing. With the aid of government subsidies, landowners are presently allowed to cut up to one-third of the trees on their land. Landsat imagery is an effective means of enforcing the terms of such agreements and preventing the overharvesting of timber.

Although Landsat's food-watch program is still a long way from perfection, it has taken us giant leaps toward controlling the production and distribution of food—preventing overproduction and the waste that accompanies it while insuring adequate food supplies in those areas where drought, malnutrition, and starvation have long been life-threatening problems.

Perhaps one day Landsat will eliminate food shortages entirely. Combined with high-powered computers and the minds of creative researchers, the ability already exists to do so.

Another important service provided by earth resources technology satellites (ERTS) is routine imaging of snow-covered ground and mountains. Accurate predictions of runoff are critical in planning the best use of water for power generation, field irrigation, flood control, and water supply estimates for major centers of population.

In 1977, Landsat accurately predicted the drought in the American West after photographs taken in February showed the snow-barren Sierra Nevada mountain range. Similar photos taken in 1975 revealed a snow line some 2,000 feet (610 m) higher.

Orbiting space stations are another valuable source of data. Such stations as America's Skylab and the Soviet's Salyut are capable of observing, analyzing, and transmitting to earth vast quantities of valuable data.

In the Soviet program, six multizonal cameras cover identical areas of the earth in different parts of the light spectrum. Once the films have been processed, the images are projected one upon another using special equipment. One film brings out details of the soil structure, including moisture content and rock composition. Another shows the types of vegetation, such as forests and plants, on cultivated land. A third concentrates on water quality in lakes and oceans and shows the extent of pollution.

The space scan, according to the Soviets, is so effective that as much information can be obtained in five minutes as a conventional aerial survey could gather in two years.

This multizonal system of photography works so well that one photograph of the rugged Soviet Pamir-Alai region obtained from *Soyuz 22* shows the glacier Fedchenko in such detail that it's possible to identify more than 100 lesser glaciers in the area, whereas only 30 were known before.

Multizonal photography is beneficial to the Soviet and world fishing industry, as well. The movement of fish and other marine life depends on the temperature of the water and the concentration and distribution of different fish species. Multizonal photography can quickly track currents of warm and cold water and predict where certain species of fish are likely to be found. It would take tens of thousands of research ships years to obtain the same information on a worldwide scale.

Some day soon, fishing trawlers may be linked to satellites, enabling them to follow great schools of migratory fish like tuna and salmon and increase catches while lowering costs. Such a union could revolutionize the fishing industry and bring new hope to the world's hungry.

An overhead view of Skylab space station cluster in Earth orbit as photographed from the Skylab 4 Command and Service Modules.

CHAPTER SIX
OCEAN WATCH

Late in the day on September 8, 1978, the fishing trawler *Captain Cosmo*, working the waters of Georges Bank off the New England coast, reported 15- to 20-foot (4.5- to 6-m) waves and high winds. A weather analyst at the National Meteorological Center in Camp Springs, Maryland, had only this single report indicating a storm. With no additional evidence, he used only the data available from surrounding points. His resulting forecast misplaced the center of the developing storm and underestimated its intensity. On September 9, the trawler disappeared.

By 7:00 P.M. that day, the storm was fully developed. The National Meteorological Center's forecast for the region predicted winds of about 30 miles (48 km) an hour and 10-foot (3-m) waves. With that information in hand, the captain of the British luxury liner *Queen Elizabeth II* approached from the east and held his ship to its course.

Early in the morning of September 11, the *QE II* encountered 70-mile (112-km)-per-hour winds and 39-foot (13-m) waves. Twenty passengers were injured, and the ship sustained more than fifty thousand dollars in damage as bow rails were bent like pretzels and hardened steel hull plates were ripped apart. The Georges Bank storm was an example of what meteorologists term "a bomb"—an oceanic extratropical cyclone whose strength intensifies explosively. Such storms are responsible for many weather-related losses of lives and ships on the high seas. Predicting them has proved difficult. The major problem is that of obtaining sufficient observational data.

In this case, the ocean-observation satellite Seasat was making a pass over the Georges Bank region at the time the storm was developing. However, Seasat's mission was strictly experimental. There was no way to process its data rapidly enough for use by weather analysts. When the data were analyzed more than a year later, they showed the developing storm in all its intensity. Had the data been available at the time, they might have saved the trawler, *Captain Cosmo*; and the *QE II* could have turned south in time to avoid sustaining injuries and damage to the ship.

Research by orbiting satellites of the seas all around us is leading to observation systems in which different satellites collect different types of data from land and sea surfaces to allow better use of the earth's resources. These satellites with their cameras and scientific instruments trained on the oceans of the world are important to us for several reasons:

- They can find and analyze sedimentation and pollution not only in oceans, but also in the Great Lakes and other large bodies of water.
- They can help detect the amount of chlorophyll in sea algae to help fishing fleets predict productive harvest areas of the oceans, thus insuring a continuing supply of an important world food source.
- They can aid in the routing of ocean-going ships, making use of ocean currents to decrease fuel consumption while cutting transit time and costs.
- They can measure and analyze the force of waves for use in the design and safe operation of offshore structures like oil rigs.
- They can map polar ice caps, ocean temperatures, and winds for improved climate and weather forecasting.

Satellites obtain data on our oceans in two ways. Automatic ocean buoys measure air and water surface temperatures, underwater temperatures, pressure and salinity, wave heights, and the speed of surface currents. These data may then be transmitted on command to a satellite, where they may be stored before being retransmitted to a ground station for analysis and use.

Another way satellites obtain data on the oceans is via direct measurement using radar *microwave* (backscatter) techniques. In June 1978, some three months prior to the Georges Bank incidents, NASA launched *Seasat 1* from Vandenberg Air Force Base in California. Although the satellite operated for only four months, it swept over 95 percent of the world's oceans every thirty-six hours, taking more measurements in a single day than had previously been taken in decades of scientific observation.

From its 500-mile (805-km) perch in the sky, Seasat pointed five instruments at the world's oceans as it made fourteen orbits of the earth each day. The instruments penetrated fog, rain, sleet, and darkness and provided data on sea-surface temperatures, wind speed, wind direction, and the amount of water in the atmosphere. *Seasat 1*'s cameras documented ocean waves, ice fields, icebergs, ice leads (openings in the ice through which ships may pass), and sea conditions along various coastlines.

These measurements may not seem important in themselves, but when analyzed and applied to specific problems, they can be critical.

One well-known example of the relationship between ocean currents and atmospheric behavior is El Niño, a phenomenon that seriously affects the fishing industries and coastal climates of Peru and Ecuador and perhaps the world climate as well.

Every several years, El Niño begins when a large body of warm water appears off Ecuador and Peru, usually around Christmas. During the year that follows, it spreads westward along the equator, producing massive rains on tropical islands that normally have light rainfall. The fisheries of Ecuador and Peru lose an entire year of

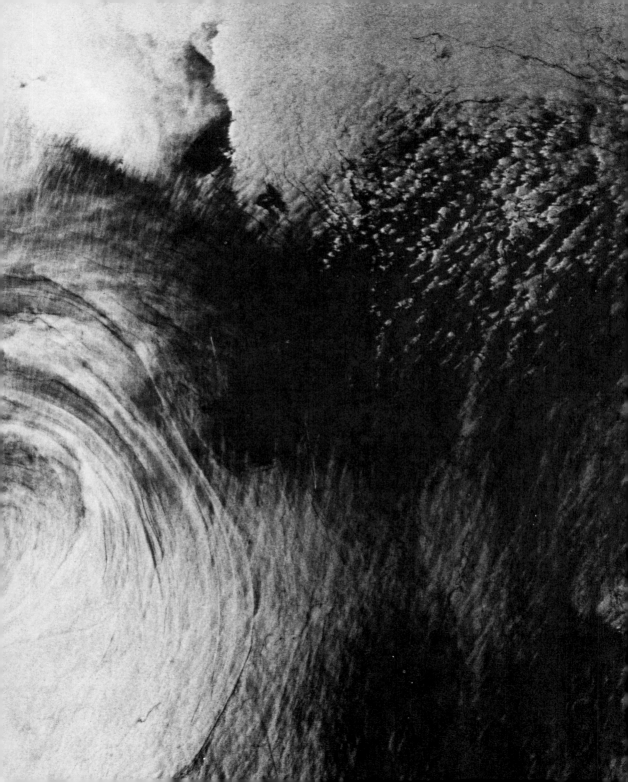

newly hatched fish because the young larvae cannot tolerate the abnormally warm ocean waters.

The economic impact of El Niño on the Peruvian anchovy fishing industry, formerly the largest in the world, is felt worldwide. Anchovy meal is a commonly used chicken food. When anchovy meal is less available, corn is used in its place, raising the price of feed corn and consequently of beef.

Recently, the region was in the midst of the strongest El Niño since 1957–58. The effects spread to the California coast, where the ocean was noticeably warmer.

There is another correlation with El Niño. The years of its occurrence have produced the most severe winters in the eastern United States. Weather analysts now know that El Niño can occur when the winds in the Pacific stop blowing easterly along the equator in October and November of the preceding year.

If present hypotheses about the origins of El Niño are correct, analysts should be able to predict this phenomenon by measuring the speed of the winds along the Pacific equator with special equipment. The drop-off of easterly winds and the development of strong westerly winds should signal its coming, allowing analysts to predict six months in advance a severe winter over the eastern United States—as well as a poor fishing harvest in South America, allowing time for the fishing industry to adjust to the coming slump.

Sea-watch satellites also aid Pacific Ocean fishermen who use satellite-derived information on the location of thermal boundaries where salmon and albacore tend to congregate because of the high nutrient levels in the waters. NASA reports the combination of fuel savings and "additional catch advantages" afforded by satellite

Seasat ocean-mapping satellites scan the waters rather than the land.

data as having an estimated annual payoff of more than $2 million.

Additional profit derived from sea-watch satellites comes from information on the ever-changing positions of the Gulf Stream and Gulf Loop currents. As early as 1975, seven oil tankers owned by Exxon Corporation used satellite-acquired data to ride the northbound currents of the Gulf Stream and avoid the currents on southbound voyages. The fuel savings for the company's fleet of fifteen tankers totaled more than a quarter million dollars in one year. The oil company now uses satellite data for all its vessels plying the east-coast route.

Sensing instruments in deep-water surveys can penetrate clear water surfaces to a depth of more than 65 feet (20 m). In the Caribbean, Landsats have recently charted shallow underwater features such as previously unknown coral reefs that can rip the hull of a ship to shreds in a matter of seconds.

But sea-watch satellites do more than scan open waters for hazards that might adversely affect humans. They also scan ice— for mariners, second in importance only to wind data.

The area of the oceans covered by ice each year is about four times the area of the continental United States. Most sea ice is moving, pushed by winds, currents, and neighboring ice floes. Some ice floes move as much as 31 miles (50 km) a day, and frequent observations of their positions are critical. Pressure ridges build up, rubble fields form, navigable lanes open and close, and travel through ice fields becomes extremely hazardous.

Adding to the difficulty of ice observations are long polar nights, clouds, fog, and blowing and drifting snow. In the past, ice surveillance was done mostly by aircraft. But today much of the reconnais-

A Seasat image of part of the shoreline of New York and New Jersey. New York City can be seen in the upper left corner.

sance is done via satellites using radar sensors, saving tens of millions of dollars a year.

The Alaskan crab-fishing industry has experienced losses as high as sixty thousand dollars a day when operations were suspended because of adverse weather. By making use of satellite-generated data, the crab fisheries could shift operations to other areas well in advance of the coming of inclement weather.

Data on sea-ice coverage can be obtained by infrared (IR) sensors aboard satellites under cloud-free conditions. When clouds prevent the use of IR, passive microwave radiometry (PMR) is used to make the necessary recordings.

The Soviet Union currently relies on four types of satellites working in conjunction with the Soviet nuclear-powered icebreaker *Sibir* to open up safe, economical routes for shipping in the Northern Seas.

Cosmos 1000, a Russian navigation satellite, provides the information used by *Sibir*'s computers to fix an exact position. Soviet Meteor satellites then give the ship's crew pictures of cloud cover and forecasts of snow and sea ice, allowing them to choose the best route. Molniya satellites allow the ship to maintain regular contact with base stations, while Ekran, the Soviet television satellite in *geostationary orbit*, sends the crew entertainment from Moscow to fill their long, lonely nights.

One of the most dramatic photos of an iceberg ever taken was obtained by *Landsat 1* as it passed over Antarctica on January 31, 1977. The shoe-shaped iceberg, nearly the size of Rhode Island, lay in shallow water, temporarily grounded north of Ross Island.

The massive iceberg was first noticed in weather-satellite photographs taken in 1971, and a search of earlier files showed that it appeared in satellite photos as early as March 1967. The photos showed its source to be an ice tongue that projected from the Princess Martha coast of Antarctica. A formation still seen on some maps but in fact no longer there, the ice tongue broke free of the ice shelf either by the force of winds or by a collision with an iceberg.

Over the years, the iceberg traveled 1,800 miles (2,896 km) along the coast, a potential menace to ships and coastline installations. As orbiting satellites continued to monitor its progress, the iceberg rammed into the Larsen Ice Shelf in August 1975, breaking off another huge iceberg nearly 36 miles (58 km) long. The original berg was grounded near the tip of the Palmer Peninsula before finally drifting slowly out of Antarctic waters, where warmer climates caused it slowly to disappear.

Some scientists believe that within the next few decades we will be able to predict short-range climate patterns based on research satellite observations as accurately as forecasters now call the shots for tomorrow's weather.

If so, then government, industry, and even individuals could use the intervening time to prepare themselves for what might come.

"The economic benefits of climatic forecasting could be very great," according to Robert H. Stewart, a research scientist at Caltech's Jet Propulsion Laboratory in Pasadena.

"If farmers in the Midwest knew that this summer was going to be hot and dry or cool and wet, they could plant crops best suited to those conditions. Or if they knew that the Russians were going to be short on wheat because of a poor growing season there . . . they might decide to do things differently."

By appropriate planning, many scientists feel, American farmers could change the economic gain or loss of the nation's agricultural output by 1 percent. That may not sound like much, but 1 percent of a $10 billion industry amounts to $100 million. And that's plenty.

Even more important, knowledge of future weather patterns could lead to extraordinary action by entire nations.

"The drought in the Sahel region of Africa, the El Niño, and other disasters of recent years," says Stewart, "have cost governments dearly. If a country knew that something like that was coming and would last only a year, it might get by by calling upon other nations for food and supplies. But if it knew a climatic event was going to last a decade, then perhaps it might have to do something else—like relocating the people in the area to be affected."

The expanding role played by orbiting research satellites grows larger every day. Satellites of more than a dozen different nations keep a watchful eye on the amount, makeup, and sources of air and water pollution. The tanker captain who discreetly dumps his tanks while anchored in coastal waters can no longer escape the watchful eye of Seasat, just as the manufacturing plant that emits harmless water vapor from its stacks by day and toxic fumes by night can no longer escape the infrared imaging of Landsat.

By diligently studying satellite-generated imagery and pursuing the source of pollutants on land and sea, we may yet be able to take our greatest step toward restoring the earth's environment.

CHAPTER SEVEN
GROWING LARGER EVERY DAY

Fueled by success and optimistic about the future, the United States and the world community of nations are continually exploring the outer reaches of our solar system and beyond. Recent satellites have probed the planets Venus, Mars, Mercury, and Uranus, while an exploration of the outer planets of the solar system began in 1972 with the launch of *Pioneer 10*, which has explored Jupiter, Saturn, Uranus, and Neptune during its *fly-by* mission. *Pioneer 10* took nearly twenty-one months to travel the 600 million miles (1,000 million km) to Jupiter. By 1987, the satellite is scheduled to have flown past Pluto on its way to becoming the first artificial object to leave the solar system. In eight million years, it will reach a point in space where the constellation Aldebaran is today!

Satellites have paved the way for many of our most notable achievements: the first landing on the moon on July 20, 1969; the launching of the rendezvous satellite Skylab on May 14, 1973; the successful deployment of the Soviet Salyut space stations in April, 1971; the 1975 "handshake in space" that linked a Soviet and an American spacecraft in earth orbit; and the launching of the first U.S. space shuttle on April 12, 1981.

Ultimately, experiments conducted in the relative weightlessness and near-zero gravity of space may lead to a whole new industry—the space factory—in which silicon chips used in computers, fiber optics used in the carrying of various electronic signals, and pharmaceuticals used in the treatment of various illnesses and diseases are manufactured to higher tolerances and greater purities than are possible on earth. These products could account for billions of dollars a year in sales and have an overwhelming effect on life as we know it.

Eventually, when we peer up into what looks to be the emptiness of space, we will actually be looking at a whole new world filled with exciting, beneficial advances to humankind—a world of nonpolluting power stations capable of transmitting energy back to earth for use in our homes, cars, and factories; of lunar space stations colonized by scientists and workers producing products from locally obtained raw materials in a new spaceborne industry; of giant space cities—entirely self-contained habitats where people live, work, and play, growing their own food, conducting valuable research, and producing products for transport back to earth and to other space colonies.

Of course, our leap from earth to space has not been without its perils. On January 27, 1967, the crew of the first Apollo mission—Virgil I. Grissom, Edward H. White II, and Roger Chaffee—died when a fire broke out in their *command module* during a launch-pad rehearsal at Kennedy Space Center. The Soviet Union suffered similar losses with their *Soyuz 1* and *11*.

Nearly 20 years later, the seven-member crew of the *Challenger* space shuttle tragically perished on January 28, 1986, in an explosion only eighty-seven seconds into flight from Cape Canaveral, Florida.

Although the loss of these ten American pioneers was tragic, they died doing what they wanted to do—what they had to do—reaching out to the heavens where humans have never gone, opening up a new frontier—perhaps the last frontier—where humans are destined to go. Like all true heroes, they died so that others might live better in future worlds.

Someday soon we may find ourselves traveling between planets with the ease that each of us today enjoys in traveling from house to school and back. Some of us may pilot interplanetary space vehicles around the galaxy. Some may make fantastic discoveries that lead to the opening up of whole new worlds. There is no end to where research satellites may eventually lead us.

None of this is likely to happen within the next few years. But nothing is beyond the realm of reality.

And it all started with the dream of a Soviet schoolmaster who decided that humans could, indeed, place an artificial satellite into orbit around the earth.

Where will it end?

GLOSSARY

Attitude control—means of turning and maintaining a spacecraft in the required position as indicated by its sensors.
Ballistic(s)—science that deals with the motion, behavior, appearance, or modification of missiles.
Command module—spacecraft compartment containing the crew and main controls.
Density—amount of matter per unit of volume.
Earth orbit—path of a body—like a satellite—acted upon by the earth's gravity.
Fly-by—space flight past a heavenly body without orbiting.
Geostationary orbit—circular orbit in which a satellite moves from west to east at such velocity as to remain fixed.
ICBM—intercontinental ballistic missile.
Infrared radiation—electromagnetic radiation of wavelengths between 7500 A—the limit of the visible light spectrum at the red end—and centimetric radio waves.
Interplanetary probe—unmanned instrumented spacecraft capable of reaching the planets.
IR—Infrared.
Lift-off—start of a rocket's flight from its launch pad.
Lunar—of or having to do with the moon.
Maria—dark lava plains on the moon once thought to be seas.
Microwaves—radio waves having wavelengths of less than 8 inches (20 cm).

Multistage rocket—rocket having two or more stages that operate in succession, each being discarded as its job is finished.
NASA—National Aeronautics and Space Administration.
NOAA—National Oceanic and Atmospheric Administration.
Planet—satellite or star; the only known planets are those of the sun, but others have been detected on nonobservational grounds around some of the nearer stars.
Probe—unmanned instrumented vehicle sent into space to gather information.
Satellite—natural or artificial body moving around a celestial body.
Sensor—electronic device for measuring or indicating a direction or movement.
Solar cell—cell that converts sunlight into electrical energy.
Solid-propellant—liquid or solid substance burned in a rocket for the purpose of producing thrust.
Tracking station—station set up to track an object through the atmosphere or space, usually by means of radar or radio.
Trajectory—flight path of a projectile, missile, rocket, or satellite.
USAF—United States Air Force.
Van Allen radiation belt—zone of high radiation density girdling the earth, named after James A. Van Allen, who instrumented the satellite *Explorer 1*.
Weightlessness—state experienced in a ballistic trajectory when, because the gravitational attraction is opposed by equal and opposite inertial forces, a body experiences no mechanical stress.

INDEX

A-4 (V-2) rocket, 10
Advanced TIROS-N satellites, 30
Advanced Very High Resolution Radiometer (AVHRR), 28
Advanced vidicon cameras (AVCs), 26, 28
Agriculture, Department of, 45
Apollo 9, 37
Atlas rocket, 14, 17
Attitude control, 39

Bureau for Aeronautics, 10

Chaffee, Roger, 64
Challenger space shuttle, 64
Cosmos 1000, 60
Crops, 45–48, 61

Defense, Department of, 11, 13
Density, 11, 22
Dream of the Earth and the Sky (Tsiolkovski), 9

Drought, 47, 49, 61

Earth, observations of, 37–44, 62
Earth resources technology satellites (ERTS), 38–48
Ekran, 60
El Niño, 55, 57, 61
ESSA (Environmental Science Service Administration), 28
Explorer 1, 13–14
Explorer satellites, 13–14, 34

Fishing industry, 49, 51, 54, 55, 57, 60
Food-watch, 45–48, 61
Fuel savings, 54, 57, 58

General Electric Corp., 38
Geophysicists, 14
Georges Bank storm, 53–54
Geostationary orbit, 60
Goddard, Robert H., 10
Grissom, Virgil I., 64

Hitler, Adolf, 10
Hurricanes, 31–33

ICBM (intercontinental ballistic missile), 13, 20
Ice surveillance, 30, 31, 54, 55, 58–61
Infrared cameras, 38
Infrared radiometers, 25, 38, 60
Interplanetary probes, 19–23
IRAS (Infrared Astronomy Satellite), 34
ITOS (improved TIROS operational satellites), 28, 30

Juno 1, 14
Juno 11, 14, 19, 22

Laika (dog), 13
Landsat imagery, 38–48, 58, 60, 62
Landsat 1, 60
Landsat 4, 39, 41, 43, 46
Livestock management, 47
Luna probes, 20–23

Mare Serenitatis, 22
Maria, 23
Mariner 2, 35
McElroy, Neil, 13
Mercury space flights, 37
Meteorological satellites, 14, 25–33, 60
Miniaturized instruments, 14
Molniya satellites, 60

Moon, 19–23, 63
Multispectral imagery, 38, 39
Multistage rocket, 11
Multizonal photography, 49

NASA (National Aeronautics and Space Administration), 25, 28, 45
National Environmental Satellite, Data, and Information Service, 31
National Environmental Satellite Service, 31
Nesmeyanov, Alexander N., 11
NOAA (National Oceanic and Atmospheric Administration), 28, 31, 45

Oceanic extratropical cyclone, 53–54
Ocean-watch satellites, 53–62

Passive microwave radiometry (PMR), 60
Pioneer probes, 19–20, 22, 23, 63
Planets, 63
Pollution, 48, 54, 62
Probes, interplanetary, 19–23

R.7 rocket, 13
Radar microwave (backscatter) technique, 55
RAND (Research and Development) group, 11

Sahel region, 47, 61
Salyut, 42, 49, 63
Satellites
    development of, 9–17
    earth, observations of, 37–44, 62
    food-watch, 45–48, 61
    interplanetary probes, 19–23
    meteorological, 14, 25–33, 60
    ocean observations, 53–62
Search-and-rescue (SAR) antennas, 30
Seasat, 54–56, 59, 62
Sensors, 38, 39, 60
Skylab, 49, 50, 63
Solar cells, 14
Solid-propellant rockets, 22
Soviets, 11, 13, 14, 17, 20, 22–23, 30, 42, 49, 60, 63, 64
Soyuz 1, 64
Soyuz 11, 64
Soyuz 22, 49
Space factory, 63–64
Space shuttle, 63, 64
Space stations, 49, 50
Spectral energy, 38
Sputnik 1, 11–13
Sputnik 2, 13
SS-6 Sapwood ICBM, 20
Stewart, Robert H., 61

Thematic mapper, 39, 41
Thor-Able rocket, 19

Thor rocket, 14, 17
Timber, 48
TIROS (television and infrared observation satellites), 25–29
TIROS-N satellites, 28, 30
Tracking and Data Relay Satellite System (TDRSS), 40
Tracking stations, 13
Trajectory, 20, 22
Tsiolkovski, Konstantin Eduardovich, 9–10, 65

United States Air Force, 31
United States Navy, 10–11

V-2 rocket, 10
Valier, Max, 10
Van Allen radiation belt, 20
Vanguard 1, 14, 17
Vanguard 2, 16, 25
Vanguard satellites, 14, 15
Vertical temperature profile radiometer (VTPR), 30
Very High Resolution Radiometer (VHRR), 30
Von Braun, Wernher, 10, 13
Voyager 2, 35

Weather, 14, 25–33, 53–54, 61
Weightlessness, 13, 63
White, Edward H., II, 64
Winkler, Johannes, 10
World Peace Conference (1953), 11
Wright brothers, 9

ABOUT THE AUTHOR

D. J. Herda has written many First Books for Franklin Watts, including two holiday books: *Christmas* and *Halloween*, and three books in the Computer Awareness First Book series: *Computer Maintenance*, *Computer Peripherals*, and *Microcomputers*. Mr. Herda lives in Wisconsin.